A HANDBOOK ON

CORPORATE
SOCIAL RESPONSIBILITY

Dr Mahesh Thakur

First Published in March 2023

ISBN: 978-93-5611-400-5

BLUEROSE PUBLISHERS

www.BlueRoseONE.com
info@bluerosepublishers.com
+91 8882 898 898

Cover Design:
Muskan Sachdeva

Typographic Design:
Pooja Sharma

Distributed by: BlueRose, Amazon, Flipkart

Contents

Chapter 1

Introduction

As widespread as the notion of Corporate Social Responsibility (hereafter CSR) has become, it is still quite a task to define what it is. And it is so for a number of reasons. For one, it is quite impossible to point out which activities qualify as socially responsible and consequently which ones don't. Secondly, anything pertaining to 'society' and 'responsibility' bear within them, the concept of subjectivity. One cannot get two people in a room to come to a definite conclusion on what is a 'socially responsible' activity, let alone CSR. Attempts, however, have been made, time and again, in defining it. What one can do is, classify the like-meaning definitions into categories to get the gist of what every category is putting their emphasis on.

Definitions of CSR

The most thorough definition of CSR considers three actors and their intertwined relationships: the state, its citizens, and multinational enterprises. Another description is the interaction between a corporation and the community in which it operates. The third broad category considers a corporation's interaction with its consumers, employees, investors, and suppliers. These are the three major categories into which we can divide the definitions, with each representing a different aspect of CSR.

However, the definition of CSR isn't the only issue where scholars and writers remain divided. In a corresponding discourse, the scope of control to be exercised over corporations in the name of social responsibility is being questioned and whether the ethos of citizens needs a revaluation before the corporations roll out their socially responsible functions. This is where the discourse takes a slight turn to insinuate the existence of a social contract between the corporations and the society, that those corporations need to be socially responsible towards.

This social contract connotes a humanitarian behavior as opposed to a self-serving one. A fair amount of authoritative works, like those of Utilitarian scholars like Jeremy Bentham, Locke and J.S Mills; or even Adam Smith's free market economy, on a fundamental level advocate for putting the needs of individuals over those of the collective society. Commonly referred to as Citizenship, the principle central to social responsibility is the social contract between the society and its actors. This social responsibility doesn't limit itself to the requirements of the present day society and its stakeholders. It includes responsibility towards the future generations that will form part of the society. As for the environment, its preservation is implicit in the responsibility towards the future.

While there is still no agreed definition of CSR, the EU Commission in 2002 defined CSR as a concept where corporations, on a voluntary basis, incorporate social and environmental concerns in their business operations and in their interactions with their stakeholders.

Corporations, Society and Discourse

A number of scholars and writers are now putting special emphasis on insisting that corporations must actively take part in the society and that the former is inherently a part of the latter. Such claims were first made in the 1970s and since then have only increased in numbers. Consequently, scholars who take a different view from this have also made their presence known in no uncertain terms. In 1973, Herrington raised a concern over how long the stakeholders of a corporation would be willing to sit through a slew of corporate non-profit activity "which appreciably reduces either dividends or the market performance of the stock."

Indeed, the importance of examining a company's social performance has not always been accepted and has been the topic of much dispute.

Scholars like Dahl, Caroll, Balabanis, Phillips, and Lyal have taken an opposing view to Herrington. Dahl has advocated for corporations to be thought of as "social enterprise", while Caroll has put emphasis on how business incorporates various expectations including ethical ones that society has from it. Balabanis, Philips and Lyal have mirrored these arguments with emphasis on how in the late 90s the mass media has put extra pressure on corporations to actively take part in the welfare of the society.

Much like Herrington, Milton Friedman and Drucker have taken a view against CSR with the former strictly ascribing the "social responsibility" of a corporation to it using its resources to make a profit for the stakeholders without having to resort to malpractices like fraud and deception. Drucker has laid emphasis on how business already turns social problems into an economic opportunity "into human competence into well paid jobs into wealth."

Moir, and Robertson and Nicholson stand on a middle ground when it comes to this discourse. Moir (2001) says that *a business's social responsibilities depend on the economic perspective adopted by the firm while Robertson and Nicholson in 1996 wrote that even statements about social responsibility by the managers can, to a certain extent, sheath a corporation from taking necessary social responsibility action in actuality.*

Dimensions of CSR

The primary goal of corporate social responsibility is to manage company activities and create a sustainable corporate culture in five dimensions:

- **Good Governance**

Good governance is a crucial factor, and its socially responsible development is absolutely required for firms that want to advance in CSR management.

Good governance needs a certain set of rules, principles, and procedures that govern the structure and functioning of organisations, promote transparency and communication with stakeholders, provide security in the management of financial and non-financial risks, and provide them with a long-term strategy aimed at socially responsible risk management.

- **Economic**

The economic part of CSR encompasses any aspect of an organization's supply chain, customer, consumer, or user management, and the socially responsible development of fiscal activities. A responsible supply chain is critical to every firm, as poor management has a negative influence on its environmental and social impact, as well as its corporate reputation, which is one of today's most valuable intangibles and one of the most difficult to revalue. Customer, consumer, and user management, on the other hand, helps to enhance the quality of products, services, and internal processes, stimulates innovation, and contributes to the discovery of new market niches, among other benefits. Finally, fiscal discipline is critical to increasing the beneficial economic and social impact on the communities where they operate, as well as in society as a whole.

- **Social**

CSR was formerly confused with social action or philanthropy, but as this article shows, CSR goes much further. In reality, social action is a subset of one of CSR's five dimensions: the social dimension. The social component encompasses activities carried out by organisations in partnership with other entities or on their own initiative that have a positive impact on society and the communities in which they operate.

- **Labour**

The labour dimension is the internal component of CSR, and it focuses on ensuring socially responsible management of the individuals who comprise the organisation. It aims to protect their health and safety, encourage them to balance work, family, and personal life, and support their professional development. It also takes into account their right to union representation, policies that advance universal accessibility, and measures that ensure equality and diversity.

- **Environmental**

Our society's relationship with the environment is currently going through a particularly delicate phase. There are other obstacles we must overcome in this field, including the climate emergency. combating the use of plastic garbage biodiversity decline, etc. Environmental management helps firms overcome these obstacles while also lowering costs, improving competitiveness, and even keeping and attracting talent, among many other advantages.

Affecting The External Environment

With the increased adoption of corporate social responsibility (CSR) initiatives, it is critical to understand how they affect various stakeholders such as employees, consumers, investors, suppliers, and the government. Employees are critical to any consideration of the origins and effects of CSR. Most micro CSR research, however, focuses on external stakeholders such as consumers and investors, ignoring employees as a critical and fundamental stakeholder group.

It is needless to say that any action/function that a corporation undertakes directly affects itself and also the external environment it functions in. This external environment includes the business environment of the corporation, the environment of the local society where it is based, and a broader global environment. This effect exhibits itself in a number of ways. The most common effect of a corporation's organizational activities includes creating job opportunities in the locale where it resides; or distributing wealth within the organization either in the form of dividends for the owners or in the form of wages for the employees.

However, the most recent concerns about these effects have been along the lines of the resultant exploitation of the environment. This exploitation could result from the use of natural resources as part of production; or the altering of the landscape for either obtaining raw materials or for storing industrial waste. An even greater concern regarding corporate activities and the environment is the emission of greenhouse gases.

As is clear, the effects of a corporation's organizational activities can not only be immense but also stretch from affecting local societal factors to global climatic factors too. One can also not ignore the fact that whether these activities are detrimental or beneficial, is highly subjective and dependent on the context in which you view them.

The Three Principles Of CSR

As has been established above, there is no universally accepted definition of CSR. Hence, in order to be able to have a standard blueprint of what CSR should entail, scholars have zeroed down three principles which make up any CSR activity.

- **Sustainability**

The term "corporate social responsibility," or "CSR," typically refers to a business's commitment to uphold social and environmental sustainability and to act as responsible stewards of the environments in which they do business.

Because it implied a duty to society and the next generation that went beyond what was required by the legally enforceable obligations of business, some corporations and economists rejected the concept of CSR. However, most businesses now use some form of CSR.

Sustainability refers to the effect of actions taken by a corporation reflecting positively on the environment in the future. Should the resources that are available in the present get depleted to the extent that they are no longer available, the activities that caused such depletion will be termed unsustainable activities. This depletion of resources becomes a greater concern if the resources are finite in nature i.e. they cannot be replenished, for example coal, oil and petroleum. Once these resources get depleted, the world must turn to alternative resources. However, this depletion won't happen in the immediate future, in the meantime, the scarcity of these finite resources keeps driving their prices up in turn increasing the operational cost of the organization. A sustainable process, in this sense, refers to using these finite and semi finite resources in a way that either individuals or corporations don't use them any more than what can be replenished. To put it simply the resources of the ecosystem should be consumed while simultaneously making arrangements for their regeneration.

Thus, a corporation, when looked at as a part of the holistic society and economy, must take measures to ensure the availability of the same resources in the future that it is utilizing now. Not only will it help reduce the cost of production in the present time, it will also ensure the longevity of the business itself. Unsustainable activities of an organization can be replaced by developing sustainable options. Or it could prepare for a future without the resources that are required in their operations now. For the most part, organizations tend to gravitate towards the former approach of sustainable operations by maximizing the utilization of the resources and cutting down on any unnecessary processes resulting in a waste of resources.

Case Study – The Collapse Of Barings Bank

Barings Bank was a global organisation based in the United Kingdom that was established in 1762. There was even a record of the English Queen. Nick Leeson was employed by Barings in 1989 and was successful there. In Singapore, where he traded on the Singapore International Monetary Exchange, he was swiftly elevated to the trading floor and became manager (SIMEX). Leeson was a brash trader who made significant gains in speculative trading. His earnings made to almost 10% of Barings' overall earnings in 1993. Because of his competence and apparent infallibility, his superiors in London did not provide him much guidance.

On Leeson's watch, a new Barings employee experienced a minor loss in July 1992. Leeson hid the loss in an error account because he did not want to jeopardise his reputation for infallibility or his job. Through speculative trading, Leeson attempted to recover the loss, but this resulted in even greater losses, which were once more buried in this account. He continued to double his stakes in an effort to recover from his losses. Later, Leeson admitted, "[I] wanted to shout from the rooftops...this is what the situation is, there are massive losses, I want to stop. But for some reason you're unable to do it. ... I had this catastrophic secret which was burning up inside me—yet...I simply couldn't open my mouth and say, 'I've lost millions and millions of pounds.'"

Leeson placed a short-term, heavily leveraged wager on the Japanese Nikkei index. His loss was so great that he could no longer conceal it when the index fell at the same time as a devastating earthquake in Kobe, Japan. A 233-year-old bank named Barings failed over night and was purchased by ING for £1. Leeson eluded capture by fleeing to Malaysia, Thailand, and ultimately Germany, where he was apprehended and sent back to Singapore. He entered a guilty plea to two counts of defrauding the SIMEX and misleading bank auditors (including the use of forged documents). Due to a colon cancer diagnosis and subsequent survival, Leeson only served four years of a six and a half year sentence in jail in Singapore.

- **Accountability**

Accountability refers to the acknowledgment of the effects of the organizational operations, whether internal or external, by the organization. These effects are quantified for the purpose of presenting them to the parties that may have been affected by the actions of the organization. The primary purpose of this acknowledgment is to show that the organization not only cares about its owners, but also takes responsibility for its actions outside of its corporate ambit. However, there's more to accountability than just taking responsibility. The corporation must make it known that the external stakeholders i.e suppliers, government, customers, trade unions etc. have a say in whether an action or the way the action is carried out, is justified.

In this way, accountability warrants that the corporation would take such measures that would not be adverse to the environment and would ensure smooth performance. The development of such performance needs to be then recorded and presented. The report must essentially make sure that the language must be accessible to the parties i.e. the report must be lucid in what it is conveying; the report must be accurate and reliable; the report must not be irrelevant to its users; the report must follow a consistent format for the user to be able to compare similar reports over time from the same organization or from different organizations.

However, it is infinitely difficult to be able to have such a report, since most reports can't rely too much on numeric data and have to, at one point, provide subjective insights into these numbers. Consequently, these subjective insights make it extremely difficult to compare such reports across organizations.

- **Transparency**

Transparency compels us to inquire about your firm in-depth inquiries like:

✓ Is your business up forward about the effects of its donations?

✓ Is your business openly disseminating the lessons it has learnt from previous rounds of corporate giving?

✓ Does your business contribute to the improvement of the corporate giving industry? Why not?

Transparency fundamentally implies that the effects of an organization's actions on the external stakeholders and the environment are not veiled in the reports of the organization. It becomes even more relevant to the external users who are not privy to the internal workings of the organizations, unlike the internal stakeholders. Much like the principles of sustainability and accountability, transparency is an equally significant process of ensuring that a corporation takes responsibility for its organizational activities and its effects, thus giving the external stakeholders a say in the process.

Gaining your customers' trust and loyalty requires being transparent. These days, your business needs to present itself as genuine, open, and purposeful in order to accomplish that. When it comes to corporate giving, the landscape has changed for businesses just like it would in any other market situation.

Investors and consumers simply demand better and more information from businesses on their CSR and corporate giving initiatives.

Conclusion

CSR is an all-encompassing topic, with no quantifiable definitions and a number of discourses surrounding it. The activities of any organization affect a wide spectrum of other organizations and individuals and even the environment. This makes it extremely difficult to narrow down the concept of CSR in just one chapter. Further in the book, these issues will be dealt with, one at a time.

Chapter 2

Externalizing Costs and Benefits

Having briefly discussed the three principles of CSR, i.e. Sustainability, Accountability and Transparency, that lay down the basis for measuring the extent to which a corporation is and should be responsible towards its stakeholders it has an impact over, we'll now look at these principles and what they entail in closer detail. The chapter focuses on how the externalizing of costs and benefits compromises the principles of CSR.

The Need For CSR

The concept of CSR has made an immense impact not only over academic and business circles, but also in the daily life of anyone that engages with a corporation at different levels, even the end consumer. This rising interest in CSR has emerged over the years as a result of poor behavioral practices towards customers, an unhealthy corporate environment for the employees, and the lack of acknowledgment of the adverse result of organizational activity on the environment. There are many other factors which have contributed to the growth of discourse around CSR, one of the most prominent of them being the general awareness about climate change and global warming. Carbon footprint has been the talk of the town for the past four years. The intelligentsia of the society have taken to thrifting clothes and adopting minimalist lifestyles in the hopes of reducing their carbon footprint and leading a more sustainable life.

Another issue has given the CSR the prominence it has today i.e. accountability from the supply chain of a business. It is no more a defence to say that a company itself doesn't engage in unethical practices like child labour or underpaid labour and that it is in fact their suppliers that do so. The corporations can no longer plead innocence by claiming they have no control over the workings of their suppliers. A fair amount of renowned retail companies have acknowledged this issue and taken steps to correct the same albeit with a heavy PR campaign. It is notable that such an acknowledgement from a corporation has indeed yielded the desired result of increased popularity of the said corporation.

Shift In The Focus of Companies

Accountability and Transparency

It is interesting to note that corporations are no more interested in a marketing campaign solely to present a responsible front while maintaining unethical practices behind doors. It has been well established now that end consumers are looking for a way to live more sustainably and thus gravitate towards businesses that are socially responsible. One must take into account how these end consumers might also be working or will eventually work for businesses, bringing with them the principles of CSR into these corporations. Thus apart from the external pressure from various non organizational actors, changes in corporations are now being seen as a result of internal actions i.e. individual effort made by its employees and managers. In this manner companies are able to fulfill two of the major principles of CSR i.e. accountability, acknowledging the problem and transparency, by being as open about their processes as their policy allows.

Sustainability

The third and one of the most difficult principles to tackle remains sustainability. The term sustainability has become a part of our daily vernacular, be it at home or at work. A lot many businesses claim that they have adopted sustainable practices in their organizational activities. However most of such claims are just another attempt to get on the wagon of greenwashing. Time and again big retail brands have been found manipulating data with vague sustainability claims in the hopes of building a positive relationship with the end consumer. This only goes to show how much positive impact a company can have if it, in reality, has adopted sustainability as its core principle.

As discussed in the previous chapter, it is difficult to determine which activities qualify as socially responsible and which ones don't. Thus the question naturally arises, how does one identify socially responsible businesses? There is a general consensus that adopting CSR means an increased accountability towards internal and external stakeholders, adopting sustainable products and processes and the likes. However these measures also lead to conflicting situations involving the expectations of the stakeholders, the reporting of such activities to the people concerned and so on. Thus, we'll discuss the problems and solutions of CSR in theory and practice further in this book.

Accounting And Externalized Costs and Benefits

Any organizational activity that has an impact over the environment, either positive or negative, is not indicated in the traditional accounting of the firm. Thus these costs get externalized, for example a factory may dump industrial waste into a river without having to pay for it. The disposal of waste which should've been part of the accounting of the company instead doesn't get included in it. And since these actions don't get included in the traditional accounting, the firm, consequently, is also excluded from the responsibility towards the environment. While cost and benefits both are imposed on the environment, it is suggested that the externalized costs outweigh the externalized benefits.

There are two major ways of externalizing costs: spatial externalization and temporal externalization.

Spatial: Spatial, as the name suggests, relates to externalizing costs in terms of 'space' or 'physical environment'. In spatial externalizing, costs are transferred to physical entities of the environment like disposing of waste in a river, or polluting the air, clearing forests etc. In temporal externalization, the present costs of the firm are deferred to the future, for example using up finite resources without providing renewable resources to be used in the future or not disposing off of waste and leaving the future managers to find a solution to it.

Temporal: When it comes to temporal externalization, the deferring of such responsibilities might not be considered much of a problem at that particular time. For example greenhouse gasses weren't recognised as a problem until much later when the depletion of ozone layer was found to be a direct consequence of greenhouse gas emission.

Conclusion

As can be seen, if such externalized costs and benefits had to be taken into account by the company in its traditional accounting, the decisions of the said company might vary significantly, and consequently organizational actions undertaken would also vary.

Chapter 3

The Stakeholder Theory

Jean-Jacques Rousseau, in the year 1762, introduced the Social Contract Theory. This theory expounded on the relationship between the individual, its society and the government. In it, he explained that the individual, by their own accord, gives up some of their rights, so that the government can make decisions for the greater good. In recent years, this theory, in particular, has gained a lot of traction in the context of CSR. This is called the Stakeholder Theory, which essentially means, that much like the individual, who gives up his rights for the greater good of the society he lives in, the company must also fulfill certain obligations, for the benefit of the society, to earn its place in the same.

Who is a Stakeholder?

This is again a contended topic, however some of the most common definitions of stakeholder state something like this: a cardinal part of the organization without the support of which the organization would not be able to exist or; any person or group who is affected by any goals or objectives that the organization achieves. Some of the most common stakeholders in an organization are managers, customers, employees, investors, shareholders and suppliers. If talking about the general stakeholders, then those include the society, the government, and the local community where the organization is based.

While many argue that only 'people' or a 'group of people' can be a stakeholder, there is also a contending view that the environment is also a stakeholder in the company since, as we read before, the company externalizes costs and benefits which affects the environment, either adversely or positively. Apart from the environment, there is another entity which is considered to be a stakeholder is the 'future' as we read previously that temporal externalization affects the future thus putting pressure on it to solve the problems being created by the present.

Types of Stakeholders

While stakeholders can be a multiple number of people, they can be classified into two broad categories.

- Internal and external stakeholders and;
- Voluntary and involuntary stakeholders

Internal stakeholders are those that function within the organization eg. the employees, the managers etc; while the external stakeholders are those that function outside the organization like suppliers and the customers.

Voluntary stakeholders have the choice of being a part of the organization for eg. an employee can resign from their position when they wish to; while an involuntary stakeholder doesn't have this choice for eg. the environment or the local society or the future.

The Stakeholder Theory

The stakeholder theory asserts that in the process of benefitting the shareholders by maximizing their profits, and consequently minimizing the benefits of the stakeholders. This theory states that while making a decision all the stakeholders of the company must be taken into account,

1. For ethical and moral reasons;

2. For benefiting the shareholders;

3. To reflect upon the workings of the organization.

Under this theory, corporate social responsibility or stakeholder management is a means rather than an end, to improve the economic performance of the company in the longer run. This view has been clearly advocated by Atkinson, Waterhouse and Wells, however it is not in concurrence with the ethical reason of the theory.

A cardinal part of the stakeholder theory remains that it attempts to identify those, to whom the corporation may be responsible while engaging in its organizational tasks. While Argenti(1993) and others have criticized the theory for failing in this fundamental task, many attempts have been made to categorize the stakeholders as discussed before. It was Clarkson(1995) who gave the two categories of voluntary and involuntary stakeholders. Another framework was devised by Mitchell, Agle and Wood(1997) to identify and rank the stakeholders on the basis of power, legitimacy and urgency. A powerful stakeholder would be placed higher in ranks and command immediate attention.

The division of stakeholders by Clarkson into voluntary and involuntary stakeholders led to a discourse where the question of the choice of involuntary stakeholders arose. Since involuntary stakeholders do not have a choice to decline to participate in the organizational activities, it has been argued that they must be given some protection from the government in the form of regulations or policies.

The Need of Data

Once it has been decided who is a stakeholder, naturally the next step should be to decide the extent to which the organization is responsible to them. In order to determine the extent of liability of the organization it is imperative to collect the requisite information for it. However when it comes to CSR, data collection is not as objective, rather it is subjective in nature, one of the many reasons why externalizing costs is excluded from the traditional accounting of the company. Two of the best performance measurements which measure objective data as well as subject data are the "balanced scorecard" and the "service profit chain".

Since every organization is different, and a single measure cannot be applied universally, a "Catalog of Measure" has been released by Cranfield University, which includes around 200 measures suited for many different organizations to measure their stakeholder management. While there are other measures which only use a single metric to measure, in comparison, multi metric measures like service profit chain and balanced scorecard are better equipped to measure something as diverse as stakeholder management.

Regulations Imposed On a Business

Every country has its own regulatory policies towards corporations. These regulations can work both ways. First would be imposing a set of guidelines on the corporations to take certain actions that may affect the external environment. Second would be preventing a corporation from taking certain actions that could negatively impact the external environment. These prescriptions and proscriptions on corporations are imposed by a country's national as well as the local government. These regulatory practices are changing by every passing year, as increasingly alarming scientific reports are brought to light regarding the harmful effects of organizational activities that the external environment is being subjected to.

One of the many features of such regulations is that the company has to make regular reports of such activities i.e. the activities of managing the stakeholders. Keeping an account of such activities is not only for the purpose of reporting to the government but also for the benefit of the managers, who in the future, can have a referential record for stakeholder management in that company.

Since the extent of such regulation is only expected to increase in the future. With such an increase, it is expected of certain proactive organizations to expand their reporting to prepare ahead for the increase in regulations in the future.

Here at this point it would be safe to presume that this disclosure of facts and reporting of organizational activities is a signifier of the growing power of the stakeholders, especially of the involuntary stakeholders who have no interests in or legal ownership of the firm.

Identifying and Managing Risks

A common topic when it comes to running a successful business is risks and how to handle them. When it comes to a business that has been mindful of its CSR and has been reporting on its activities due to the regulations, the factor of risks increases. There's an intricate relationship between sustainability and risks, however, most corporations have a limited understanding of what sustainability or sustainable development is. This limited understanding in turn allows the corporations to plead innocence when it comes to calculating environmental risks. As a result reports and management of risks is deficient in a number of ways.

To be able to have a holistic understanding of environmental costs and benefits for the purpose of investment analysis, a company must first identify the types of costs and benefits that are needed for such evaluation. Once that is done, it becomes fairly easy to quantify qualitative costs and benefits. This identification, collection and quantification of data allows the company to see these costs and benefits in financial terms and hence be better equipped at managing risks.

A lot many benefits of computing this data in this way are not reflected immediately. For example, an enhanced company image, better relationship with the local society, and the regulators which in turn can speed up the approval of new projects that a company overtakes. These are the kind of benefits that can only be realized well into the future which brings us to an important factor while computing the data for costs and benefits. While identifying the benefits, the companies are advised to incorporate the time that it will take to redeem these benefits.

Conclusion

The Stakeholder Theory is a breakthrough in the study of managing an organization, since it takes a holistic approach towards solving this problem rather than looking at CSR as something that a company should deal with individually in isolation.

Chapter 4

Sustainability

While all three principles of the CSR are of extreme importance, the principle of sustainability is essentially important because of how much traction it has gained in past few years not only in the business world but also and quite especially in the normal world where people are making conscious efforts towards "slow fashion", sustainable lifestyles and the like. However we need to consider what sustainability means in terms of corporations and their organizational activities, and for that we need to, once again, look closely at the definition of sustainability.

The Brundtland Report, officially named 'Our Common Future', gave the foundational definition of sustainability, from which every other definition has been derived. This report was based on sustainable development and thus defined sustainable development, after which the standard definition of the same became, "Development which meets the needs of the present without compromising the ability of the future generations to meet their own needs." This definition or principle of sustainable development has been widely adopted as one of the foundational principles of many reports, committees, commissions etc., like Agenda 21, Maastricht and Amsterdam Treaties, the Rio Declaration.

The Brundtland Report and Its Drawbacks

What is the Brundtland Report?

This report gave us strategic measures for sustainable development. Along with these strategic measures the report put emphasis on how our technology and societal makeup is hindering the world's ability to move forward and have a better future. The report works on the premise that sustainable development is possible which has led the world to a discourse on how to achieve this goal. There is a fair amount of confusion over what sustainability is and if it is any different from sustainable development. Most scholars view both the terms synonymously since the former usually implies development in a sustainable way and such a development is taken to mean that status quo is maintained with the only difference being that sustainability is now a part of the scenario. When we talk about sustainability in the business world, this theory of maintaining the status quo is our prime concern. If in the corporate setting, achieving sustainable development while maintaining the status quo is attainable, then this sustainability is actually an added benefit rather than it being a part of the bigger sustainable picture.

Drawbacks of The Brundtland Report

The three main concerns with this report are:

- It assumes that sustainable development is preferable and is also possible. And that sustainability means for a company to be able to exist in the future;

- It works on the assumption that sustainability and sustainable development are one and the same thing;

- A sustainable business can exist by only working on environmental and social issues.

A major undisputed assumption about sustainability in the Brundtland Report is that a company will be able to grow and that only the economics of development has to be considered in order for such growth to happen. The issue of the terms sustainability and sustainable development being synonymous stretches back to the management lingo that has been used for over thirty years. In management terms the word sustainable implicates a company's ability to simply be able to continue to exist in the future. In addition to this understanding of what sustainability means in the corporate world, there are more reports that work on the same assumptions as the Brundtland Report. A major issue with all these reports is that they presume that sustainability and sustainable development, both, can be achieved merely by addressing the Triple Bottom Line i.e. social, economic and environmental aspects of performance. To be able to better understand the difference and the preferred outcomes of sustainability and sustainable development, we should replace these two terms with the word 'durability'.

When the cost of capital is considered, while investing in a business, a safer company, with a low cost of capital is considered to be the best choice. One of the factors that determine whether a company is safe or not is whether it is sustainable, in the corporate sense. Whether that company will exist 20 years from the moment of investment. This is where the real purpose and meaning of sustainability gets lost. A lot many oil extracting companies have started using the term sustainability in their reports and visions. Most of such companies have renamed certain terms like from "oil" to "energy". While focusing on renewable energy for the future is a part of sustainability, the fact that these companies are still majorly involved in the processes of extracting oil, points out how very far away their actions are from their vision of sustainability. It is not enough to be able to continue as a business in the future and call it sustainability.

Reimagining Sustainability

It is time that we move away or at least attempt to add to the definition of sustainability as given by the Brundtland Report.

The elements of sustainability should include the following:

- **Society's Influence:** to indicate the kind and extent of impact that the society can make on a company as discussed in the previous chapter.

- **Effect on the Environment:** to measure the effect a corporation and its organizational activities can have over the geophysical environment.

- **Organizational Relationships:** to study the relationship between the internal stakeholders of the company.

- **Financial Monitoring:** determining suitable returns when there is risk involved.

While these present a holistic view to achieving sustainability and sustainable development, a balance of all four elements is necessary. An overdrive of one element while neglecting the other, will lead to failure in realizing the goal of sustainability in corporations.

Another thing to keep in mind while handling sustainability is how the effects, whether good or bad, will be distributed among the stakeholders of the firm. It is of equal importance that the stakeholders be on the receiving end of the effects of sustainability as much as owners are, and such effects must be affected in a manner that is acceptable to the stakeholders.

Conclusion

Sustainability over the past few years has been used synonymously with many terms that are loosely associated with practices that are deemed to be harmless or conducive to the future and the environment. However, the term durability is better suited to replace the terms sustainability and sustainable development. Durability has broadly speaking two major elements. Equity and efficiency. Efficiency needs to be thought of in terms of environment and not finance. Equity needs to ensure that, like owners and shareholders, the stakeholders too deserve to receive whatever consequences are affected by such durability.

Chapter 5

Ethics and CSR

Ethics in business is quite a common topic. While ethics and corporate behavior may vary from country to country, it is universal in the sense that the corporate world in every country has a standard set of rules and regulations to be followed by the companies and its employees. When these standard sets of rules and regulations are seen as a responsibility, these are then known as the ethical code of conduct of business. The idea behind ethics in business is to behave decently in every organizational activity. Businesses in reality usually study and implement ethics in situations like a conflict of interest which might compromise the profits of the firm, as the primary purpose of ethics. Issues of social responsibility are treated as the secondary purpose of ethics.

Business Ethics stands somewhere along the lines of honesty, integrity, respect, differentiating the good from the bad. Ethics in general and when it comes to business is quite vague, the boundaries are not defined and depend a lot on subjective moral standards and judgment. Because of this subjective nature, business ethics can be studied through many a different approaches and philosophies. Aras(2006) discussed how failing to adhere to business ethics, might lead to a company failing to achieve its goals.

Philosophies in Ethics

The very importance that ethics is now given in the corporate world is enough of an indicator of how much the societal environment or the stakeholders have influenced it. Another factor that has contributed would be the lack of ethics in organizational activities previously seen in cases where there were mass casualties.

Ethics is a highly debatable topic. What might be ethical to one person might be unethical to another, even if the two disagreeing persons belong to the same party. It is a highly fertile ground for dispute and it would do us some good to have a look at the various philosophies of ethics.

Deontological Ethics: It is believed here that some actions are inherently wrong and thus such actions should be considered ethically wrong without any qualifications. The issue with this philosophy however remains on how it is to be determined which actions are wrong or right. For example, killing a person is morally wrong, however in circumstances of a war, such an action is permitted, even encouraged.

Teleological Ethics: Here our main focus remains on "the right" and the "the good". "The right" is seen in the context of "the good" and how the former helps maximize the latter. Here the ultimate result is what determines the right or the wrong, as opposed to the actions which have led to those results.

Utilitarianism: the premise of utilitarianism is based upon the idea that the larger good for the society is the result of when the individual members of the society pursue things and act for their own interest. A few theorists later added governments as mediaries to better help achieve this goal. Here too the outcomes are given more importance.

Relativism: Relativism simply denies that there is anything that can be considered universally true.

What might be right for one side might be wrong for the other. Thus ethical relativism advocates that there are no moral rights or wrongs that can be universally acknowledged without any amendments to them. One can divide ethical relativism into two sub categories:

- *Conventional*: According to this subcategory, the standards and morals of the society we live in, or the corporation exists in decides the ethical code of conduct one must adhere to.

- *Subjectivism*: Here we look at our personal scale of what we considered morally right and wrong.

When it comes to conventionalism, a major problem arises because we must determine where our ethical loyalties lie. There are many groups we can belong to at the same time which may have conflicting ethical values. We could be a part of a cultural society, but also a part of a professional society whose ethics and morals may be at odds with each other.

Objectivism: This philosophy stands in sharp contrast to relativism as it advocates that while some groups may be at odds with each other in matters of ethics, there are certain universally acceptable moral standards and principles that do not change throughout these groups, whether or not they are acknowledged.

The Gaia Hypothesis, Corporate Behavior and CSR

To understand the increasing popularity of ethics in the corporate realm, we must understand the Gaia hypothesis and how it has brought a new outlook to corporate behavior. In the late 1970s James Lovelock came up with his Gaia Hypothesis where he proposed that the earth sustains life because it is a self regulating planet. According to this hypothesis, every being that makes up this earth is interdependent on each other and every action taken by any one being, inevitably affects others. Up until very recently, in the economic sphere, the idea of classical liberalism was prevalent. Wherein, it's every creature for themselves and their actions either do not affect the other elements or the effect is limited to only to the elements which are intended to be on the receiving end of the action. When applied to the corporate world, the Gaia hypothesis works in sharp contrast to this classical liberalism that has been observed in the economic sphere for a long period of time. While before the firms only had to focus on their actions, the advent of the concept of CSR and the Gaia hypothesis that emerged around the same time, has pushed the corporate world to pay attention to the effects that organizational actions can have outside of the firm regardless of whether or not such an effect is directly related to the firm.

Corporate behavior is behavior characterized by many different parameters but most primarily by legal, ethical and socially responsible ones. The effect of corporate behavior extends not only to the shareholders and stakeholders but also to the entirety of the economy. An ethical, socially responsible and of course legal behavior by a corporation in any event, be it strategic planning or taking decisions, ensures that the actions of the firm remain sustainable and subsequently help in the long term survival of the firm.

For a company to be socially responsible, it must realize that it is a part of the society and that every action that it takes has a direct and indirect effect on the society, of which its shareholders and stakeholders are a part. Such a realization should help the company in monitoring its corporate behavior and altering this behavior in a manner that is consistent with the ethical code of conduct of the society and the company. An ethical corporate behavior lays a solid groundwork for CSR in the corporation.

This would be an apt place to mention that there is a growing importance of the concept of corporate reputation. An investor is better placed when making an investment in a company with a good corporate reputation. It is needless to say that building a reputation takes a while, while having it shattered only takes a single blow. However, empirical evidence suggests that having a good corporate reputation not only ensures a healthy and trustworthy relationship between the stakeholders and the corporation but also helps in shielding the firm in times when it is being criticized or is under scrutiny.

Conclusion

Ethics and CSR go hand in hand. As we have seen, according to the Gaia hypothesis every action of any entity in an ecosphere has a direct or indirect consequence on another entity. In the light of this knowledge, corporations must assume responsibility for their corporate behavior and attempt to modify it to better suit the needs and requirements of the stakeholders instead of only looking out for the business owners. Such modifications in the corporate behavior should be made within the realms of the ethical code of conduct of the company and the society. An ethical corporate behavior observed over a period of time can help build a corporate reputation for a company which is not only desirable but also has shown positive results when it comes to the longevity of a company even when faced with turbulent events.

Chapter 6

Performance

When it comes to the management of organizations the performance of the organization is dependent on the evaluation of the performance, and its reporting. When it comes to CSR such performance evaluation and reporting is equally important. However, as discussed earlier, the collection of data and subsequently measuring it in matters of CSR becomes difficult through the traditional methods. In this chapter we'll discuss the evaluation and reporting of performance.

Whether an organization's performance has been good or bad is highly subjective and would depend on the perspective of different stakeholders. Thus determining a standard set of indicators to indicate what good performance entails is imperative when it comes to measuring the performance. Measuring financial performance is easier than measuring stakeholder performance since the measurement of the former is based on objective scales. Since objective measures of stakeholder performance are not available, there are a few standard subjective measures we could apply according to an annual survey published in

the journal Management Today. These measures
are:

1. Marketing Quality
2. Innovation Capacity
3. Management Quality
4. Quality of goods and services
5. Ability to retain high performing employees
6. Community and environmental responsibility
7. Use of assets
8. Financially sound
9. Providing long term value as an investment

Social Performance

The 1970s saw a rising prominence of the concept of social performance. This was largely a result of how the performance of the businesses in areas other than the stock market or value creation for the shareholders was being scrutinized. This performance came to be known as the social performance of a business. To be able to measure social performance, social accounting was developed which measured aspects like concerns of the investors, focus on community relations, performance in relation to ecology, impact on geophysical environment etc.

Many scholars believe that it is not essential to give reasons as to why social performance should be measured, however Solomon has mentioned not only a reason i.e. to help businesses in making informed and rational decisions, but also has proposed a model for such accounting. In his model he has proposed aspects which could be measured to determine social performance however, he hasn't made any mention of how those aspects are to be measured. This is one of the biggest deficits when it comes to calculating external costs and benefits and subsequently the social performance.

Methods of Measuring Social Performance

Churchman: Performance measurement and evaluation is an important aspect in the management of any organization. The components necessary to measure in social performance, according to Churchman, are specifying objects to which the result will apply to, the language in which the report will be presented, a standardized method to measure so that the test can be applied to different organizations and accuracy.

Kaplan and Norton: In 1992 Kaplan and Norton came up with a new perspective on performance measurement. They contested the traditional accounting methods since they tend to have a control bias towards financial function while measuring. They proposed what they called a Balanced Scorecard and contended that this scorecard makes strategy and vision as the central focus. The four components of the balanced scorecard are:

1. Financial Perspective: relates to the perspective that the shareholders have of the firm;

2. Customer Perspective: relates to the perception of the customers of the firm;

3. Internal Business Perspective: relates to the integral function of the firm which the firm must command excellence in;

4. Innovation and Learning Perspective: relates to measuring the capacity and ability of the firm to create value in the future.

They argue that this scorecard helps in keeping track and measuring the long term and short term performances of the firm. Moreover every business that chooses to use the balanced scorecard will be able to measure performance even if the specifics of the firms differ. The balanced scorecard provides what they consider are the four core aspects of performance to be measured: mission, strategy, technology and culture and hence gives managers a comprehensive four perspective look into the performance. Since it takes into account the perspectives of two stakeholders i.e. the customer and the shareholder, separately, Kaplan and Norton suggest that the companies should tweak the scorecards according to the relevance of certain stakeholders or certain objective goals the firm may have.

Measuring Effects on Environment

It is imperative for an organization to understand the effects that its organizational activities are having over the environment before it starts developing measures to regulate those effects. To be able to understand such effects of the organization, the organization needs to undertake an environmental audit. ISO 14000, a family of standards that helps organizations in regulating their actions which have a negative effect on the environment, is universally related to environmental audits. These environmental audits must observe the following issues:

1. Regulatory compliance;
2. Effectiveness of measures taken to reduce pollution;
3. Current state of energy usage and the possibilities to become energy efficient;
4. Current state of waste disposal and the possibilities of waste reduction in the future;
5. Current state of usage of sustainable methods and future possibilities of developing renewable resources;
6. Life cycle of products;
7. How capital investment might affect such issues;

As can be seen an environmental audit requires a comprehensive knowledge about the processes of the organization, from the very beginning to the end and the possible outcomes of such processes in the future. A traditional company accountant cannot undertake such a heavy task that requires the involvement of specialists. The objective here is to ensure that there is a comprehensive understanding of the effects of organizational activities before jumping to evaluate them. An environmental audit also implores the company to look into places that it may have either ignored in the past or had been unaware of.

Since conducting an environmental audit can be a magnanimous task, it is advisable that instead of conducting it as a separate isolated process, the issues discussed above should be incorporated and addressed in the normal decision making of an organization as a contiguous process.

Measuring Performance

When measuring performance, two issues need to be considered to get a comprehensive and accurate measurement: the reason for such measurement and what needs to be measured in order to satisfy the first issue. It is of utmost importance that the purpose of the measurement of performance be clear in order to evaluate performance with greater precision. While the reasons for measuring performance are many, but largely it is done to help the organization in strategising better, having greater and informed control in its action and to be accountable for its actions. According to the measurement theory, measurement is a comparative process where comparisons are made in the following areas:

1. Temporal, where the comparison is made through different periods of time;

2. Geospatial, where comparison is made with another business or sector or country;

3. Strategical, where comparison is made with the subsequent result of different procedures or course of actions.

Not only is the measurement of the performance relative, but also performance itself is. It doesn't exist in isolation. Whenever a business is presented with a report whether financial or otherwise, the numbers are looked at in comparison with the projected and the eventual outcome, or with past reports and statistics. Social performance is no different. Since such a comparative study has to be done when it comes to performance, it is essential that there is a quantitative approach when it comes to measuring the qualitative aspects of the performance.

Evaluating Performance

Measurement of anything is effective only as far as it gets evaluated. It must be addressed at this point that no measurement can be applied universally to every organization since they can have widely differing goals they wish to achieve from performance measurement. To be able to avoid using measures that aren't well suited to the needs of an organization, evaluating the measures of the performance is an essential part of the measurement itself. Since there are many reasons for measuring performance, during the time of evaluation it could lead to differences between different departments of the said organization. Thus managers are expected to have the ability to resolve or even avoid such conflicts arising within the organization by communicating information in an effective manner so that no stakeholder feels neglected by the performance measurement.

Another issue to be kept in mind while evaluating performance is performance sustainability. It must be ensured by the evaluators that the current performance can be sustained and the implications that it will bear in the future are recognized. To be able measure future implications, measures should be devised that can project future performance as closely as possible. This will help in assessing the performance in a comparative manner as discussed before.

Conclusion

While many other factors go into managing an organization, social and environmental performance measurement and evaluation go a long way in ensuring that there is an informed and better decision making process within the organization, the resources are allocated with a view of achieving greater sustainability, organizational activities are undertaken with a greater accountability towards the impact they may bear on the external environment.

Chapter 7

Globalization, NFPs and CSR

Globalization has taken the world by the storm. From affecting major corporations down to the streets of third world countries, everything we see has been touched, even if remotely, by globalization. One of the major effects that it has had on the economy is the increase in competition between businesses. This increase in competition has been advocated to give the consumers choice to choose between different businesses but has also made the corporation driven by profit maximization. Such a profit oriented mindset can harm the corporation in its way to becoming sustainable. Since corporations have become such an integral part of the society, it is important that we address the issues and solutions regarding the socially responsible behavior of a firm in the context of globalization.

While the traditional definition of the term globalization implies that it is an exchange of goods, services, ideas and capital. This is a definition that limits the actual effect that globalization has on any country. With every passing day the world is brought closer through the internet. Websites, apps and softwares that can be accessed from any part of the world. Apart from the extreme economic and political effects it has had, it has also affected society, religious philosophy profoundly. Such effects had led to drastic changes in countries and such changes could potentially lead to undermining the effectual power that a government has over the functioning of a country. Naturally policies have been made by governments of various countries in order to protect its economy and various other aspects of state regulation. What concerns us is the question as to what happens to a company and its CSR in the light of governments bringing these policies.

Why Globalization is Essential to the Study of CSR

There have been a swirling amount of changes that globalization has brought about. Globalization has led the world to be more fast paced where businesses are required to cater to the needs of their customers not only as quickly as possible but also offer a cheaper price. It has led to an increase in competition among businesses and in turn significantly transformed the consumer's behavior. Consumers are now looking for instant gratification for a cheaper price. To be able to cater to the needs of its customers, businesses have had to resort to a number of unsustainable practices, for example, offering competitive salaries to employees or cutting down costs in the manufacturing process, often by resorting to unsustainable processes, in order to reduce the cost of the end product.

Another concern is that since globalization has opened up the local markets of a country to global businesses, there are many international standards that a company has to adhere to, whether those are standards pertaining to the manufacturing process or the end product itself. These standards could also pertain to accounting and auditing of a global firm. With the regulation of international standards, there comes a threat of deregulation of local markets and their assimilation into one big global market. Another major effect of globalization is the susceptibility of markets and businesses across the globe getting affected by any financial crisis happening in just one part of the world.

However our issue remains as to how globalization is essential to the study of corporate social responsibility. Many theorists over the course of the 20th century have argued that while globalization does affect the economy in various ways, it also affects the fabric of the society. Rapid change in economic growth and development cannot happen in isolation while the society remains untouched. Another issue arising out of the effect of globalization on CSR is the question of responsibility to safeguard the rights of the society and its individual. Should the government look over the welfare of the people or the corporation, and if the responsibilities have to be divided, in what ratio are they to be divided. As per the current scenario, the government makes laws and regulations to regulate the market while the corporations have control over the commercial conduct of businesses. According to older norms, the corporations merely had a responsibility towards the shareholders by increasing its profits to keep the shareholders interested. However, now the corporations have to take into account, as previously discussed, the sentiments and the well being of the stakeholders, have stricter ethical standards they are held to, and their performances are evaluated or are at least encouraged to be measured and evaluated keeping these factors in mind. The companies have also realized that the shareholders intend to retain their positions as shareholders for a long period of time. This intention of the shareholders makes the company look towards making themselves sustainable in order to last long not only temporally but also in quality.

Corporations must face the challenge of globalization through the processes of debates and deliberations with a view to introducing international standards for business behavior. Scherer and Palazzo have argued for the need of political CSR that apart from taking into consideration the other aspects that we've dealt with in consideration of CSR till now, will also take into account political and social issues like forced labor, human rights etc.

Failures of Corporations in a Globalized World

Over the years many corporations have been involved in scandals involving misappropriation of funds, corruption, exploitation etc. It is a belief of many that incorporation of socially responsible behavior would inevitably prevent any scandals since CSR ensures longevity of the firm and consequently works to protect the firm from losses that could affect the growth of the firm negatively. As previously established, a corporation that adheres to the principles of CSR has a responsibility towards its stakeholders, which include the locality it functions in, the geophysical environment and its resources, the future, its employees etc. The role of company managers is also extremely important when it comes to managing the organization guided by the principles of CSR. A company already in control of its responsibilities towards these stakeholders is a lot less likely to commit acts of social and moral irresponsibility. As we had discussed earlier, apart from the corporations fulfilling their duties towards the stakeholders, the stakeholders too have a duty to keep the company in check through regulations. In this sense, CSR is a concept which makes it essential for both the stakeholders and the corporations to work towards the same goal of ensuring good performance by the organization in order to reduce possibilities of corporate failures in a globalized world.

The question then comes on to how these managers and stakeholders must ensure socially responsible behavior. It must be noted that socially responsible behavior is not synonymous to legally sound and abiding behavior. If we take a look at the Enron scandal of 2001, it would be quite clear to us that all the managers were well aware of the regulations and yet did not raise any concerns and neither did they take any steps to change the corporate behavior. It must be understood that stricter regulations and laws do not prevent corporations from engaging in socially irresponsible behavior. These stricter laws and regulations are strict only till such a time as someone from the organization finds a way to circumvent those regulations. Thus our solution in trying to curb such scandals, that result not only in a loss to the corporation but also cause a loss to the society, lies not in legal remedies but regulating corporate behavior in a manner that ensures sensitivity towards the society and other stakeholders. It is imperative that the firms focus on producing financial value for their shareholders but not in a manner that compromises the ethical standards of the company which could in turn undermine its CSR.

Does Globalization Interfere with CSR?

It is difficult to establish the extent to which globalization affects CSR. While in developed countries the governments have tried to maintain and regulate standards of organizational operations through laws, the condition in developing countries still remains slightly worse off. It is here that globalization leaves its worse effects since developing countries do depend on globalization for their growth, but are far too unequipped to bring effective laws against their exploitation. The lack of balance between the government and the companies often offsets many chances of avoiding adverse effects of globalization. It isn't only about the lack of balance between the government and the corporations, but also between the corporations as well. While some corporations adopt sustainable practices, others simply chase their goals of profits maximization. Globalization is a rapidly spreading phenomenon and is often left unchecked. Since there are companies that do not engage in socially responsible activities it is imperative that preventive and corrective measures be taken up through CSR. As for the preventive measures, it is essential that corporate behavior must be regulated. Ethical behavior should be a fundamental part of encouraging socially responsible behavior. The management, employees and the shareholders must be aware that unethical behavior or suborning it will lead to negative consequences. Internal accountability is necessary if a company wants to build a workforce that is ethically and socially responsible. As for the corrective measures we must take a look at the not for profit organizations.

Not For Profit Organizations

While we are on the topic of globalization and its relationship with CSR, we must briefly examine Not For Profit(NFP) organizations that have been taking up responsibilities to correct many wrongs or rather been trying to restore the wrongs that were left in the wake of socially irresponsible activities in a globalized world. These organizations, as their name suggests, are not driven by the aim of profit maximization unlike corporate organizations. An NFP fills the voids by taking care of issues that are left by the lack of regulations by the government and the lack of responsible behavior by the corporations in the wake of globalization. Because of this nature of these organizations, it is being suggested that instead of having a negating connotation of "non" or "not" in their name, such organizations should be named in a way that it is more representative of the work they do as opposed to representative of the work that *corporations* fail to do. Many organizations refer to this sector as "Citizen Sector Organization" or "Civil Society Organization."

A distinguishing feature of NFPs is that it isn't driven by profits and the criteria for making decisions naturally differs from that of for-profit corporate organizations. Another notable feature is that the people who avail the services of these organizations don't pay for those services. The money to pay for these services comes from various other sources, like donors, public funds etc. An NFP could be an organization that gets funded and is directly and completely backed by the government of the territory it works in. Or it could be partially funded by the government and partially dependent on private sponsors. It is however important to note that the demand for their services far outweigh the supply of resources in that organization. Among the many different NFPs the most notable and widespread categories of NFPs are public NFPs that are funded by and/or derive their powers directly from the government. Then there are educational institutions that provide education at all levels at a subsidized fee, while receiving their funds either from public resources or from private parties.

A very important category of NFPs is charity. Globally these form one of the larger portions of NFPs. A charity is an intra-vires organization, which essentially means that it cannot do anything that goes outside the ambit of its core purpose. To put it in context, the opposite of intra vires organization is ultra vires organization. An ultra vires organization can do all that it is not illegal for it to do even if it circumvents the primary purpose of the organization. Charities on the other hand are formed for a very specific purpose called the 'charitable purpose' and are supposed to restrict their activities around the said purpose only. Many other such restrictions are put on charities as well. Other restrictions are put on charities as well, for example politics doesn't fall within its ambit. It can raise funds, however, it is prohibited from raising funds by trading even though it is quite common for charities to trade by employing a third party. Charities have an overwhelming number of volunteers as well as employees. While having volunteers does cut down a lot of operating costs, there is a constant conflict that comes from the relationship between the employees and the volunteers.

Issues With NFPs

As we've already seen, NFPs are not regular commercial organizations. Some categories of NFPs like charities, as discussed, have special regulations upon them even though they get tax benefits. There are a wide variety of NFPs and it is needless to say that regulating them while maintaining uniform standards would be injustice to their cause. However, given that the nature of NFPs, for the most part, remains consistent throughout the spectrum of their categories, the issues that are faced by them can also be broadly classified.

Accounting

Since the goal of NFPs is not to push for profit maximization, it naturally leads to the organizations being pressed for resources. This in turn leads to the organization having to look for funds and become more stringent about allocation of funds. To this extent, the concerns of this sector end up seeming very similar to corporate organizations, for example, optimization of the process, reducing transaction cost etc. However the decision to these issues are decided very differently since the goal of both the organizations is different.

A frequent source of trouble is that the goals of NFPs almost do not have a finish line i.e. when the NFP would fulfill its purpose completely is almost never known since largely such issues are chronic and have to be worked on for over many years, sometimes decades. The problem lies in the fact that the funds that come in are not enough to be considered for a long term financial plan. Since the funds are so low, they have to be utilized for the next immediate need even if it means surpassing the long term goal since the fund is not sustainable enough to last that long.

Fundraising is something frequently undertaken by the NFPs. When it comes to purpose specific funds, for example, a disaster relief fund, the funds that are given to NFPs are to be used only for that purpose and for nothing else. If the purpose of the fund is fulfilled while the fund itself is not fully exhausted, the remaining fund cannot be used elsewhere. This becomes an issue since the fund can be used for other causes but then it would become easier to exploit funds through means of legality. It could be argued that the remaining funds could be allocated for other operational costs like paying the administration involved in, say, the disaster relief programme, however, it would again become a loophole for the NFPs to exploit through legal means, which could set precedent for using specific funds for paying the administration. It must also be noted that if the fundraising campaigns were to announce that the funds received would also be used to pay for general expenses of the NFP along with the purpose for which funds are being raised, people would be less likely to donate.

Performance

It is difficult to provide any profit incentives to the managers working for an NFP since their primary purpose is not profit maximization. Since the purpose of NFPs is a social cause, while operating with little to negligent funds and resources, one motivation to work becomes acquisition of resources and funds to fulfill those causes. Since there is no traditional motivation of profit nor the required funds for the NFPs to work, their performance tends to suffer overtime. This brings up the issue of determining the parameters of good performance for NFPs along with the standards and reporting of such performance.

Another issue with the evaluation of such performance is that the beneficiaries of such organizations hardly ever take part in its evaluations. For instance, in other commercial organizations, the consumer inevitably shows its approval or disapproval of goods or services provided by the organization by way of buying or subscribing to such goods or services. In an NFP the receiver of the services is rather a dormant participant in the process for the most part and cannot offer any substantive insight into performance evaluation.

Management

The first and the most obvious concern for the managers of an NFP is to look for resources that fuel the NFPs goals and their subsequent allocation to minimize operational cost as much as possible. Given how these organizations are pressed for funds, the planning for utilization of these resources in the longer term becomes quite futile since the managers do not know if the funds would last that long. There is also an issue of considering the potential of competition between two NFPs. The competition doesn't necessarily stem from the race to be able to reach the beneficiary before another NFP with a similar niche does, but rather from the tussle to obtain funds and resources. Dealing with issues such as these take up the primary concerns for the management of NFPs.

A problem that the management itself can pose quite a few times is when a comparatively larger NFP, which is much likely to attract additional funds, has a larger market share. Thus management of NFPs tend to work towards increasing their market share, oftentimes losing sight of the original purpose of the NFP. There is also the issue where when the purpose of an NFP is fulfilled, it must cease its operations. However, this is where the human element of the management of the NFPs creates trouble by striving to extend the lifetime of the organization even if its purpose has been achieved.

Resources

It is no surprise that the biggest issue with NFPs is their lack of resources and funds, the primary reason for every other issue that we have discussed above. As stated earlier, in quite a few places the funding for NFPs is sourced from public funds and such NFPs even derive their powers directly from a government. Educational NFPs are a prime example of public funded NFPs. However not all NFPs have access to funds from the government and have to look for other means such as fundraising and trading(through third parties or through a trading subsidiary). Borrowing funds could be another way to gather resources however, this method couldn't be adopted if the goal isn't a capital project and there is a security the NFP could offer.

The Principles of CSR and NFPs

Throughout the book, we've seen how the traditional lens of looking at corporate social responsibility assumes that an organization participating in it on papers is not necessarily indicative of it being a socially responsible corporation. It is no different when it comes to putting NFPs under scrutiny to check for socially responsible behavior. The fact that such organizations work for the benefit of the society does not exempt them from the same standards of scrutiny other commercial organizations are put under. The concern of CSR is how the organization is juggling the roles,responsibilities and rights of the stakeholders rather than fulfilling a formal obligation on papers.

Stakeholders

An NFP is most likely to have a spectrum of varied stakeholders unlike commercial organizations with similar stakeholders across different organizations. As we've already discussed previously, stakeholders form the most important part when evaluating socially responsible behavior of any organization. The division of power that a stakeholder is given in an NFP as opposed to a commercial organization also varies with the stakeholders in the former case having more power than those in the latter one. The conflicts between the stakeholders of an NFP may greatly differ from a commercial organization and consequently the resultant conflict resolution would also differ.

Accountability

It doesn't sit well if a not for profit organization chooses not to maintain accountability with its stakeholders. NFPs are not autonomous bodies that sustains itself with its own income; rather they derive their funds to run the organization not from the work they do but from other external sources like the government, donors etc. For such organizations it becomes even more imperative rather than just optional to be accountable towards the donors, beneficiaries and the society at large.

Transparency

While a private commercial organization is encouraged to be transparent, it is by no means bound to be transparent unless it wants to be socially responsible or is subjected to an audit. However when it comes to not for profit organizations transparency, again, becomes an imperative rather than a choice since the fund is not generated by the organization itself but rather is donated by various other players. It also becomes important to remain transparent to let the stakeholders know whether or not the funds that they had allocated to the organization are being used for the purpose they were donated for.

Sustainability

NFPs actually work to restore all the damage done by unsustainable activities undertaken by other profit driven organizations. They work towards ensuring the resources are distributed equitably; the resources are used in a manner that ensures their enjoyment by future generations as well; because of a lack of funds they try to work their best in only a small amount of resources. However the problem of sustainability for such organizations does not lie in whether they are making the external environment around them sustainable since this is the very goal of NFPs. Rather it is ensuring that the internal management of the organization is carried out in such a sustainable way that the organization doesn't exhaust all its resources before it even fulfills its purpose.

Conclusion

NFPs in a globalized world are doing repair work for all the damage that has been caused by profit driven organizations. However, even the NFPs are not immune to the implications of the principles of corporate social responsibility. The implications that the NFPs are subjected to and those that the profit driven organizations are subjected to are different since both are motivated by extremely different incentives.

Chapter 8

CSR and Management

We've seen how important planning everything around corporate social responsibility like responsible accounting, ethical measurement and reporting is imperative to maintain the sustainability and relevance of an organization in this day and age and for the future as well. However, it is only management strategy and effective leadership that can help achieve goals leading to a better and sustainable organization. In this chapter we'll look at how management strategy and leadership can help achieve CSR obligations of an organization.

Roles and Duties of Management

A manager in present-day commercial organizations can have many different roles either combined or vested in different managers. These roles could range from being a marketing manager to a project manager to a human resource manager etc. While the specific roles of a manager may change, the duty of the manager remains to achieve the goals set by the objectives of the organization. These roles generally take the form of the manager of an organization taking notice not only of the internal department of which he is a manager but also the external environment in which the organization functions. Keeping track and notice of the external environment which relates to the market, customers, suppliers etc., forms the strategic part of the manager's job. This fundamental strategic duty of a manager across the different roles he could have in an organization remains the same, which makes it easier to study and decipher their duties and roles in regards to CSR and management strategy. To be able to study such roles and duties, we need to broadly identify a few objectives that businesses generally have. These objectives may not stand true for all organizations, but in most cases at least one objective stands true for every other organization, if not more.

Achieving Business Objectives

To be able to study the strategic roles and duties of managers of an organization, we need to broadly identify a few objectives that businesses generally have. These objectives may not stand true for all organizations, but in most cases at least one objective stands true for every other organization, if not more.

1. Profits and Cash Flow:

Very few organizations like not for profit organizations as discussed earlier, do not aim for profit maximization. Other than that, for the most part almost every organization aims to achieve the goal of profit maximization. However, how these profits are to be achieved and what effects the process of achieving it might have on the organization in the longer run are the prerogative of the managers. The conflict arises when achieving profit maximization as a short term goal interferes with the long term goal of maintaining sustainability of the organization.While profits are what the organization earns, cash flow is what keeps the organization alive. Ensuring that there is not only a consistent cash flow but also making efforts to increase that cash flow also falls under the strategic duties of the management.

2. Maintaining the Ratio of Profit to Capital:

This could take two major forms. For organizations that are commercial in nature, return on capital employed i.e. the financial ratio which measures an organization's profitability to the efficiency of the capital invested, the managers can be expected, during their due course of duties, to compare the performance of their organization to those that have a similar niche or a similar business model in order to get a clear picture of how well the organization is performing in terms of return on capital employed.

An equivalent of return on capital employed for a not for profit organization is the ratio with which they provided a service to the capital employed. Again the manager's strategic duties include comparing how well their not for profit organization is performing relative to other similar NFPs.

3. Focus on Improving Brand Value:

While assets of and profit made by a commercial organization; assets of and funds raised by a not for profit organization add to the capital value of the organization, they do not add much to the brand value of the organization. Or at the very least these capital assets cannot replace the brand value since a temporary financial setback cannot harm the company's goodwill for too long; rather, it is often the brand value of the company that pulls it out of a financial setback. However the same cannot be applied vice versa. A huge profit cannot compensate for a company losing its brand value in the market. Even if there is a financial gain, such a practice of gaining profit at the cost of brand value is not sustainable practice leading the company to sooner or later go under.

4. Long Term Sustainable Growth:

All businesses seek to expand their assets as well as their share in the market. However, in doing so many businesses lose sight of the importance of maintaining a sustainable expansion. A rapid growth spurt in assets and market share may not last for a long time. If adjustments to the business are made without considering the fact that the growth may not last a very long time or that the current growth is inconsistent and unpredictable, it could cause the business to be unable to pick up the pace later on when the growth spurt comes to a stand still and ultimately go under. This is why it is necessary that businesses should put their focus more on sustainable, steady and organic growth rather than an unstable and unpredictable growth.

5. Affective Outcome Over Optimal Solution:

Satisficing is a decision making process wherein a decision is made based on utilizing the resources that are readily available to the organization and are adequate for the task at hand rather than striving to get the best possible outcome regardless of the cost of the process. The focus is on a sustainable process rather than the outcome.

Most organizations, whether commercial or non commercial(NFPs) will find that at some point in time or the other, one or more of these points have been or are their objectives. It is not, however, necessary that an organization which has more than one of these objectives goes out of its way to ensure that all of these objectives are given equal importance. The organization reserves the right to prioritize one objective without completely giving up another.

Functions of a Manager

The role of a manager depends on whatever title he holds in the company, his duties include achieving one or more business objectives of the organization, but along with these roles and duties there are certain functions, that can be broadly classified, that a manager has to take part in and be the deciding factor. Now that we've looked at some of the objectives that an organization can have and expect the managers to achieve these objectives, let's look at some of the other functions that a manager has to engage in.

Planning and Decision Making

When a target is set, it is but natural for a manager to go into planning. It is in this stage that the managers have to take into account different approaches to achieve the target. This is why it is extremely important to look not only for quality but also quantity. A quality plan might end up costing the organization more and thus reducing the ratio of capital employed and as discussed before, maintaining a good ratio of capital employed is one of the most important objectives of an organization. A broad classification of the process of planning involves developing a premise or in other words, taking inputs to figure out the available resources and appropriate process that should be followed to make the most out of the resources. It is important at this stage that multiple alternatives should be considered that could help in achieving the target. This is important since the output of the plan has to take into account how other departments of the organization are going to be affected. Even the premise or the input of the plan could change depending on what other departments plan to do. Planning is the stage where one can explore a wide range of possible solutions and strategies since once the process is put into action, it can be very hard to go back on the decided plan and implement something new.

To sum up the stage of planning, a manager must take into account other departments and their strategies and outcome of their plans and explore several possible solutions to narrow down on the most cost effective plan to achieve the target. Planning is primarily of two types:

1. Strategic Planning

Strategic planning relates to the strategic duties of the manager i.e. planning to fulfill the objectives of the entire business rather than just the target or task assigned to the department under the manager. This includes doing market analysis, analyzing consumer behavior, what they want, their purchasing power, their willingness to put their money into buying from the business. Once this is done, the capacity of the business to deliver on such demands is analyzed. This type of planning is more future oriented. It is important to plan strategically and study market and consumer behavior regularly in order for the business to have its objectives outlined and remain relevant in the future.

2. Corporate Planning

Corporate planning comes into play once the strategic plan has been laid out. Strategic plans lay out the broader outlines of the course of business while corporate planning puts more definition into those broader goals. The internal departments of the business are given targets and tasks they have to achieve while remaining within the ambit of the objectives laid down in strategic plans. After an external and internal analysis, multiple courses of action are laid out for the manager to pick one and make a decision on how to proceed further.

One of the primary functions a manager is involved in, in any organization, is decision making. Of all the options he may have while deciding on a course of action, it is his job to take into account all the factors, negative or positive, and make a decision. There are certain steps that all managers must take into account before making a decision. The first one includes identifying the objectives as to why the decision must be made. Which objective will the decision target? If there is one objective that needs to be given more priority over another? Questions like these help in narrowing down and subsequently identifying the purpose for which the decision has to be taken and all other objectives of the organization that could possibly be affected, either positively or negatively, through the said decision. Once the objective of the task is identified the second step is to take multiple courses of action into account to be able to decide which course of action will give the best possible outcome with the available resources. The third step is to implement the decision made on the basis of the first two steps.

Evaluation

Evaluation, it could be said, is the most important function of a manager. After a decision has been made and the plan has been put to implementation, the manager must evaluate how well the course of action was planned and how well it was executed. The manager must then figure out

- How much difference and deviance was there in the purported course of action and the executed one;
- In what all aspects of the plan did such deviation occurred;
- If the deviation could have been avoided had they taken any other decision;
- If the course of action taken because of such deviation was helpful in achieving the target;
- If the course of action caused by the deviation affected any other departments of the organization, whether positively or adversely.

Such evaluation tends to point out the weaknesses of the department and of the organization as a whole. Without such an evaluation, certain trivial and avoidable mistakes might get repeated time and again causing a significant loss to the organization overtime. Such loss may not be monetary necessarily. Such mistakes could end up causing an adverse effect on the functioning of other departments and create friction within the teams and managers; or it might ruin business relations with stakeholders outside the organization for example the consumer or the supplier. Such friction inside and outside the organization could lead to decreased efficiency in the running of business.

Leadership in CSR

It has been observed that for an organization to exhibit socially responsible behavior, it requires effective leadership. Simply because one is a manager and is seated at a position of a leader it does not make one have leadership qualities. As a leader the manager does have the power to give orders and make his subordinates do what he thinks is fit in a course of action. This is merely an administrative job that he has to fulfill as a manager. When it comes to leadership it is reflected in the process of how a plan is being executed under the manager. Leadership puts its focus more on how the manager manages to convince his subordinates and sometimes his equals to undertake a task. Thus the qualities of leadership lies more in ability to communicate and motivate. Leadership qualities could also depend on the nature of the organization or of the department. For eg. if the nature of the organization is utilitarian which requires for a task to be finished the leadership would have to be very strategic in assigning tasks in order to meet the deadline.However, styles of leadership can depend largely on three factors:

- The personal disposition of the leader;
- Those subordinate to the leader;
- The given situation at hand.

Thus assuming that leadership is a simple concept would be doing it injustice. It is a concept that is capable of shifting power dynamics in any given situation. People are drawn to leaders who exhibit leadership qualities rather than those that simply have the title of the leader and power is capable of falling into the hands of a capable leader.

Power

It is quite predictable that when talking of leadership the topic of power has to be discussed. Many writers and philosophers have attempted to define power, however, implicit in all their definitions is the idea that power is inherently present in our daily lives. When it comes to an organizational setting, power is explicit in every transaction. For eg. one department of the organization having more power over another and such a power dynamic translating into the respective manager of the powerful department having more sway in decision making over other managers. Power flows through many sources. It could come from people who believe that a certain person should have the power; or it could flow from the sheer expertise that a person may have over a subject which naturally makes the subordinate accept his power and leadership etc. The point is that power may not necessarily arise and reside in one particular person. Even if it has been established that the manager will lead a project or head a department, it does not automatically make him a leader or give him the power of one. Power does not stay in a particular position with a particular individual. A person with great expertise on a matter, even if he is a subordinate to the manager, will have more power or sway in the decision making process. The quality of leadership in a manager is tested in such situations. Will the leader allow a junior member of the team to have a significant say in the matter? If yes, then it could lead to a better functioning of the department. If not, it could lead to friction within the department. Thus simply being assigned the role of a leader will not ensure that the manager has leadership qualities.

Agency Theory of Corporate Governance

In general parlance it is assumed that the shareholders employ the managers of an organization as their agent and their primary function is to ensure that the wealth of the shareholders is maximized. Another common thought is that the managers are expected to make such decisions, as objectively as possible, as the shareholders would make in the normal course of carrying out their business i.e managers are expected to function on behalf of the owners of the business. However, this view fails to take into account the human element of managers having different motivations and directions of thought than the owners.

This is where agency theory of corporate governance comes in. This theory takes into account the fact that the agents(managers) and the principal(owner) have different vested interests in the business and these interests are entirely self-serving. Starting off on this premise helps rationalize the difference in socio-political behavior of the agent and the principal. The theory intends to make the relationship between the managers and the owners a contract of agency. In such a contract, the interests of both the parties will be kept in mind so that in the future the boundaries of both are well defined making the decision making process a lot easier and satisfactory to the managers and the owners.

However there are certain limitations to this theory as well. Firstly, it is quite unrealistic to assume that the owners of the organization would be rational about giving some of their autonomy up, for the purpose of giving the agent a little more control and benefit even if it acts as a motivation to the agent to perform better. Secondly, an organization need not be very single dimensional with just one shareholder and one manager. The agency theory may work in a small organizational setting, however, bigger organizations, which are the primary concern of CSR, may not work in that setting. Multifaceted contracts between different agents and different principals could lead to unhealthy corporate relationships.

Conclusion

As we've discussed throughout the chapter, management and leadership are an extremely important element in maintaining the corporate social responsibility of any firm since at the end of the day, it is the managers of the organization that are planning and implementing organizational activities. In this chapter we've seen that not enough for a manager to be in a leadership position, but also to inculcate leadership qualities in himself for an efficient working environment in the organization.

www.ingramcontent.com/pod-product-compliance
Lightning Source LLC
LaVergne TN
LVHW051304200726

843510LV00010B/1280